T4-AKN-605

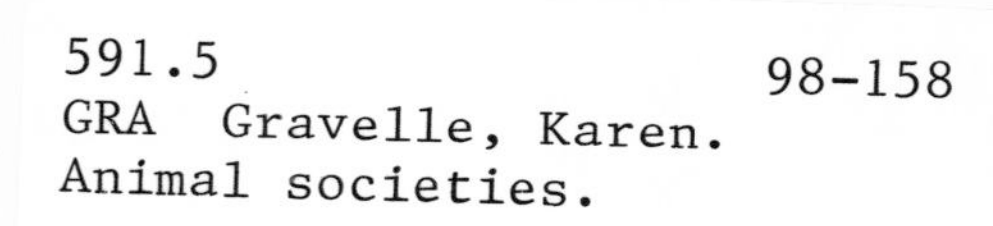

DISCARD

BCCC
a31111000981587n

ANIMAL SOCIETIES

ALSO BY KAREN GRAVELLE

LIZARDS

UNDERSTANDING BIRTH DEFECTS

ANIMAL SOCIETIES

BY KAREN GRAVELLE

A VENTURE BOOK
FRANKLIN WATTS
NEW YORK/CHICAGO/LONDON/TORONTO/SYDNEY

TO ELLEN DORSEY

Photographs copyright ©: Animals Animals: pp. 1, 2 (John Chellman), 3 (Michael Fogden/OSF), 5 (Leonard Lee Rue III), 6 top (Zig Leszczynski), 6 bottom (Ron Willocks), 9 (Jack Wilburn), 10 (Patti Murray), 11 (Stouffer Productions), 12 (J.H. Robinson), 13 (James D. Watt), 14 (Ben Osborne/OSF), 15 bottom (Mantis Wildlife Films/OSF), 16 (Raymond A. Mendez); Bettmann Archive: p. 45; Wade Sherbrooke: p. 4; Karen Gravelle: p. 7; Devin Scott: pp. 8, 15 top.

Library of Congress Cataloging-in-Publication Data

Gravelle, Karen.
Animal societies / by Karen Gravelle.
p. cm. — (A Venture book)
Includes bibliographical references and index.
Summary: Examines the societal structure among various animal species, such as wolves, sharks, penguins, ostriches, ants, and bees.
ISBN 0-531-12530-0
1. Animal societies—Juvenile literature. 2. Social behavior in animals—Juvenile literature. 3. Animal communities—Juvenile literature. [1. Animal societies. 2. Animals—Habits and behavior.] I. Title.
QL775.G68 1993
591.5′248—dc20 92-35877 CIP AC

Printed in the United States of America
6 5 4 3 2 1

Contents

ANIMAL SOCIETIES

CHAPTER ONE

Introduction: A World of Difference

Imagine that you are a Zeenon female— the mother of a four-year-old daughter and a baby son. Like other females in your society, you live in a group with your own youngsters, your female relatives, and their dependent offspring. Until last year when she died, your mother, Hoga, headed the family group. A powerful matriarch, she guided the family in almost everything they did.

Although old for a Zeenon female, Hoga was especially formidable in coordinating the family defense against enemies. Strangely, protecting themselves would seem to be easy for Zeenons. Very few other groups would risk attacking a family of these Amazons. With their intelligence, their strength, their weapons, and their undying loyalty to each other, adult female Zeenons are more than a match for most attackers. Even Zeenon males, who are much larger and stronger, would not dream of threatening one of these tightly knit groups.

But, like all youngsters, baby Zeenons are vulnerable and need protection. And, in spite of adult Zeenons' general invincibility, there is one other society, the Yorbins, that poses a deadly threat to them. With Yorbins slaughtering more and more Zeenons each year, your group will miss Hoga's long experience in avoiding death at their hands. Your oldest aunt has taken over Hoga's position as matriarch, but it remains to be seen if she will be as effective.

You look over at your little son, playing at the edge of the river. By the time he is thirteen or fourteen, he will be gone too. While your daughter will stay with the family group for the rest of her life, your son must leave as soon as he matures, for adult Zeenon males do not live with females. He may join with other males in a loose, informal group or he may live alone—most likely, he will do both over the course of his life. Since your son will probably seek mates from other female groups, the only time you are likely to encounter him as an adult will be when hundreds of Zeenons join together on their long seasonal migrations to faraway sources of food and water.

In a nearby forested area live the Xerbes. Although close relatives of the Yorbins and like them in many ways, the Xerbes have never attacked Zeenons. In fact, they never venture into the plains where the Zeenons spend much of their time, so the two societies cross paths only when Zeenons enter wooded areas.

Like the Zeenons, the Xerbes live in family groups. In their case, however, adult males live with females and hold the top leadership positions in the society. There is considerable rivalry among males for these positions, and battles between them can be fierce. Occasionally, alliances between males can determine the outcome, as when two or more lower-ranking males join together to unseat the chief. Females also jockey for the top position among themselves, and the offspring of females that attain the highest status tend to grow up to be high-ranking adults themselves. Even though Xerbes work together to find food, because of their dominance over females and young, adult males usually get first access to whatever is found.

While young Xerbe females sometimes leave the group in which they grew up to join a neighboring group, no male of any age is welcome in another group. And no Xerbe—male or female—goes off to live a solitary existence.

Like the Zeenons, the Xerbes are also skilled at group defense. In their case, however, the superior strength of the males is an important factor in whether they win or lose. Because of their smaller physique, the Xerbes have many more enemies than the Zeenons do. In addition, the Xerbes' relatives, the Yorbins, often take them captive, and even neighboring groups of Xerbes can present a lethal threat.

Xerbes establish territories where they live permanently. Each Xerbe group fights to keep rival Xerbes out of its particular area. Occasionally, in a fight over turf, the stronger of two Xerbe groups completely annihilates the other, killing every male, female, and baby. Such a horror could never occur in Zeenon society!

Have you been able to guess who the Zeenons, Yorbins, and Xerbes are? If you thought they were different groups of people, you were only one-third correct. The Yorbins, with their guns and greed for ivory, *are* human beings. Their close relatives, the Xerbes, are chimpanzees, however, and the Zeenons are elephants.

It may surprise you to know that many animals, including some that are not our close relatives, live in societies that have much in common with our own. Even Zeenon (or elephant) social organization—in which females live together in groups, visited only periodically by males—is not as nonhuman as it seems. While no such human culture exists today, an example can be found in the Amazon society of our mythological past. Whether different from or similar to ours, almost all animals live in some kind of society.

WHAT IS A SOCIETY?

If you're like most of us, you probably think of a society as a number of people living together (or at least associating with each other) as a group. This group is organized in some manner, with a set of rules that define who belongs to the group and how the members are supposed to behave toward each other. The particular organization and customs of this community are what make it different from other societies.

On an individual level, when we refer to a particular person as social, we are usually talking about someone who has a lot of interactions with others in his or her group.

Since almost all animals must mate in order to reproduce, all have some interactions with members of their own *species*. Thus, most animals are at least minimally social. The type of interactions animals of a species have with each other and the frequency with which those interactions occur determine the kind of society typical of that species.

Like different people, some species are more social than others. The more members of a species depend on each other for survival, the more social that species is said to be. In general, fish, reptiles, and amphibians depend on members of their species only for mating and, thus, they have developed relatively primitive societies. Mammals and birds, on the other hand, not only mate but raise their offspring. Therefore, their social interactions and, hence, their societies, are more complex. The most social animals of all, however, are certain ants and bees. In these species, the entire society devotes its efforts to raising the young of only one female.

There are other reasons besides raising young that prompt animals to live and work together in organized groups. For certain animals, mutual protection or food gathering is an important reason for living together. In order to understand what makes some animals form intricate societies while others spend most of their lives alone, let's take a closer look at an animal society we're all familiar with—our own.

HUMAN SOCIETIES

By definition, all mammalian newborns, including humans, are dependent upon their mothers for milk to survive. This means that all mammals begin life living with at least one nursing female. But newborn humans need others of their species for more than just food. Human babies are completely helpless at birth. Unlike the newborns of mammals such as horses, elephants, or dolphins, a human infant cannot sit up or roll over, much less follow its mother as she searches for food.

Moreover, even after young humans have learned to walk, feed themselves, and perform simple chores, they remain dependent on adults. In large part, this dependency is due to the type of intelligence characteristic of humans. While we have a unique ability to learn new things, there are some drawbacks to this kind of intelligence. Unlike many fish, reptiles, and amphibians, whose brains are preprogrammed with most of the knowledge they need to survive, we are born knowing very little. In fact, almost all the information critical to our survival has to be learned.

Thus, human children need adults to teach them the things they must know and to protect them until they have learned enough to be able to live on their own. This means that a human mother must remain with and care for her children for many years.

Humans have additional reasons for living together even after they are fully grown. With few natural weapons and little speed, our early ancestors had to work together if they were to successfully hunt larger, stronger, faster, and often fiercer animals. Even more important, by banding together, they were better able to defend themselves against the many animals that preyed on them.

Thus, group living became the key to early human survival. Because this is as true now as it was then, human societies of today share strong similarities with those of our ancestors. The basic unit upon which all human societies are built is the family, although different cultures often have different ideas of exactly what the word "family" means.

As human groups became bigger, the need for an organized structure increased. In particular, these groups needed someone to be in charge if group members were to be able to cooperate successfully. In a family, the individuals in charge are the parents. As several families banded together, the strongest, most skillful, or wisest person was chosen as the chief of the clan. The line of authority became even more complex as groups grew even larger. Thus, all human societies are hierarchical to some degree; that is, all have some line of command. As we shall see, this is one respect in which human societies have much in common with the societies of many other animals.

Finally, in order to work together, members of a group must have ways of communicating with each other. Humans use a combination of sounds, facial expressions, and body positions to send signals to each other. Although many other animals also use sight and sound signals, some rely on smells, touch, or even electrical signals to communicate. Not surprisingly, we find it much easier to communicate with animals who use messages that are similar to our own.

As we shall see in Chapter 2, the close similarity between wolf and human societies is the primary reason why dogs, and not some other animal, became "man's best friend." But first, let's look at some animal societies that are very different from our own.

Most cold-blooded vertebrates, such as fishes, reptiles, and amphibians, have evolved different strategies for survival than those used by mammals and birds. Thus, it makes sense that their social interactions and societies differ from ours too.

Instead of being dependent upon adults, most baby fishes, reptiles, and amphibians are completely self-sufficient from birth, knowing instinctively how to find food and seek shelter. Although they are able to fend for themselves, however, this does not mean that these newborns are as capable of surviving as their parents. For one thing, they are much smaller and thus more vulnerable to *predation.*

There are two ways fishes, reptiles, and amphibians can compensate for the defenselessness of their young. The first, and by far the most popular, is to have a great many offspring in the first place. This way, even if the majority die, enough will survive to become adults. While animals who must take care of their newborns are limited in the number of offspring they can handle, this isn't a problem for these parents, since their young can care for themselves.

A second way these animals can deal with the vulnerability of their offspring is to provide protection for them. But cold-blooded vertebrates are not very likely to do this. For one thing, fishes, reptiles, and amphibians have few opportunities for interacting with their own offspring. Usually, females deposit their eggs and then leave them. When the young finally emerge, their mothers are gone. In other cases, several females may lay their eggs in the same place, making it impossible to tell which offspring belong to whom.

Finally, many adult fishes, reptiles, and amphibians defend themselves against predators not by fighting but by hiding or fleeing. Thus, with the exception of a few animals, such as the alligators and crocodiles (discussed in Chapter 3), most aren't really able to come to the defense of their offspring.

As a result, in the vast majority of these species, adults have little to do with their young. Therefore, unlike human societies, the societies formed by these animals do not have parents and their offspring as the basic unit. But that doesn't mean that fishes, reptiles, and amphibians don't interact with each other. In contrast to the social interactions of group-living species, however, many of the interactions between these animals are designed to help individuals avoid each other.

A perfect example of a social organization in which many of the interactions are intended to ward off others is that of certain lizards, such as the anoles of Florida or the spiny lizards of the Southwest. In many of these species, adults spend a lot of time and energy establishing individual areas, or territories, which they claim for their own exclusive use. Having a territory is very important to these animals. By keeping others out of its immediate neighborhood, a lizard is able to reserve more of the food there for itself.

Both anoles and spiny lizards have distinct messages they use to signal other lizards to get out of their private turf. When one of these lizards discovers another trespassing on its territory, it runs over to the intruder, bobbing up and down, puffing out its sides, and extending a brightly colored area on its throat. This is a signal to the other animal that the area is already taken. Usually these strange-looking push-ups are sufficient to encourage the stranger to move on to another place. If not, the two lizards will actually fight it out, with the winner taking possession of at least part of the territory.

Both male and female adults of these species establish territories, but those of males are much larger and overlap with the territories of several females. Although females exclude other adult females from their territories and males exclude other adult males, females and males generally tolerate each other's presence, at least during mating season. Otherwise, there would soon be no lizards to talk about!

Male anoles and spiny lizards also use push-up signals to ask females to mate with them. Even baby lizards communicate with push-ups, although in their case it's not to establish a territory or attract a mate. Newborn lizards bear an unfortunate similarity in size, shape, and coloration to the large beetles and crickets their parents like to eat. Scientists think that juvenile lizards may use push-ups to alert adults to the fact that they are lizards, not insects.

Thus, the societies of anoles and spiny lizards are made up of single adults, each living on its own private piece of real estate. Although baby lizards are somewhat tolerated within these territories, they are completely ignored by their parents and must fend for themselves. Because males' territories overlap those of females, males and females may enter each other's territories.

Many male frogs also establish territories during mating season, and crocodiles defend territories around their basking sites. Even some fishes, such as male sticklebacks, chase other males away from the area around their nests. But many fishes, reptiles, and amphibians do not even engage in these territorial interactions. Hidden under the leaf litter on the forest floor or in cracks and crevices on land and underwater, they don't interact much with others of their species except during mating season.

All of us are aware of some situations, however, in which fishes, reptiles, or amphibians can be found grouped together in large numbers. For example, many fishes school together for safety, and clusters of snakes often hibernate in the same protected spot over winter. But even when packed side by side, these animals are not as social as they might seem. For the most part, actual interactions between individuals in these groups are few.

In other cases, animals may look as if they are living together when they are not. For example, since the breeding territories of some frogs are very small, a lily pond full of croaking males may appear to be one big group instead of a collection of tiny, separate—and vigorously defended—kingdoms.

Because they rely on other members of their species for very little, fishes, reptiles, and amphibians have relatively primitive societies. But this doesn't mean these societies are less useful than the more complex social organizations of some other animals. A simple organization with few interactions between individuals is the type of society best suited to help members of these species to survive. Thus, for them, the arrangement is perfect.

MAMMALS

Since humans are mammals, all of us have an idea of what at least one mammalian society is like. However, while all newborn mammals live with their mothers, how and how much they interact with each other when they grow up varies widely from species to species. In fact, many mammals are not very social at all. In more solitary species, adults—particularly males—spend most of their lives alone.

Bears provide a good example of this type of mammalian social organization. As adults, these animals have virtually no need for each other, except as mates. With their size and strength, they have little to fear from other creatures and therefore have nothing to gain by grouping together for defense. Nor do they need each other's help to obtain food.

In fact, by living together, bears would actually reduce their chances of survival. It takes a lot of food to feed an adult bear, and the presence of more than one adult in an area would make it very hard for each animal to find enough to eat. Therefore, bears space themselves out, each sticking to its own territory.

Like lizards, adult bears have ways of signaling "Get out!" to others that trespass on their private turf. Since bears have a very good sense of smell, they can tell from droppings left by another bear when a particular area is already occupied. Intruders may also be able to tell by other signs whether or not they would be likely to win a fight with the territory's owner.

In some species, male bears stand on their hind legs and make deep scratch marks with their claws high on the side of a tree. When an intruder male sees these marks, he stands on his hind legs and tries to reach higher. If he fails, he knows that the owner is bigger and is likely to outclass him in a fight. This is usually enough to encourage the trespasser to move on to a more hospitable area.

During the breeding season, males use their sense of smell to find females that are ready to mate. Once mating is finished, the two go their separate ways again.

Since raising their young takes a lot of time and energy, female mammals give birth to only a few at a time. In the case of grizzly bears, a litter consists of one to four cubs. Tucked away in a den, the cubs are nursed until they are old enough to follow their mother as she searches for food. By tagging along as she digs up roots and bulbs, locates berries and honey, fishes, and hunts small animals, the youngsters learn to forage for themselves.

Male grizzlies do not participate in raising the cubs. This is fine with females, since a mother grizzly is able to feed, teach, and protect her offspring by herself. In fact, since male grizzlies occasionally kill cubs, females with young attack any male that dares to come near.

After two years, grizzly cubs become adults and leave their mother's home range to find territories of their own. With the exception of siblings who may remain together for a while, the males will never again live with another of their species, and females will live only with their own young cubs.

Occasionally, adult grizzlies gather together at places where food is especially abundant, such as salmon streams or caribou calving areas. Although they may be in the company of other bears, they are far from friendly. Using grunts, roars, and attacks, the most powerful bears command the best feeding spots. Thus, except when mating or raising young, grizzly bear communication is designed to help individuals steer clear of each other.

Several large mammalian predators, such as tigers, cougars, and leopards, have developed societies that are similar to that of bears. On the other hand, grazing mammals, such as deer, horses, antelope, and elephants, tend to be far more social, in part because they need each other for protection. Among the most social mammals, however, are the wolves and dolphins (discussed later), whose societies have much in common with our own.

BIRDS

Birds, too, have developed a variety of social organizations. Some are similar to most mammalian societies in that only females raise the young. But in 90 percent of bird species, males take an active part in feeding and caring for their offspring. There are a number of reasons why this is so.

Among the many differences between birds and mammals is that, unlike mammals, birds do not nurse their young. This fact has had an important effect on the ways in which male and female birds relate to each other and how they evolved to care for their offspring.

Another important difference is that, while all mammalian newborns are dependent upon their mother for food, birds vary widely in this respect. Ducklings, for example, can swim and forage on their own within hours after hatching, although they still need their mother for guidance and protection. On the other hand, the *nestlings* of many other species, including tiny songbirds and majestic eagles, are helpless. Needless to say, the degree to which newborn birds are dependent on their parents also influences the type of society that a species develops.

In most avian, or bird, species with helpless young, males share in the care of their offspring. Since baby birds do not nurse, male birds—unlike mammalian fathers—are just as able to feed the young as females are. Moreover, with their high metabolic rate, baby birds need a lot of food—much more than one parent alone can supply. Thus, without the male's assistance, the nestlings are likely to die.

Because the efforts of both parents are necessary to raise their offspring, birds with helpless young are usually *monogamous*; that is, they have only one mate per breeding season. Males with more than one mate would tend to spread themselves too thin. When males concentrate their efforts on one female, their young obviously have a better chance of survival. Equally important, since females of these species need a male's exclusive help in raising their young, they are unwilling to share a mate with other females.

In these species, therefore, society consists of a male, a female, and their offspring, at least during the breeding season. When the young are grown, the parents may separate, mating with new partners the following year.

However, since raising helpless nestlings requires a lot of cooperation between a couple, pairs that have worked together successfully are usually better off picking each other as mates the next year than trying a new, untested partner. This is why some birds, such as storks and cranes, mate for life. In some other species, individuals tend to return to the area where they previously nested, thus increasing the chance that a pair that has separated over the winter will be reunited.

In contrast, mammalian societies are more likely to be

polygamous; that is, males may have more than one mate at a time. Since many male mammals do not help in feeding their young, they can afford to have more than one mate without jeopardizing their offsprings' survival.

Because the young are not disadvantaged by this, females of these species do not object to a male that has other mates. Since males in these societies compete with other males to acquire as many mates as possible, those with the most mates are the strongest and the most likely to father offspring that will survive and reproduce. Thus, females actually *prefer* them to males that have no other mates. Elephants, seals, and sea lions are good examples of these species.

For the same reasons, some male birds are also polygamous. Roosters, for example, actually have harems. Their chicks are able to feed themselves, so these males can devote their energies to acquiring as many females as possible. Roosters fight among each other, with the prize—hens—going to the strongest. Roosters that lose either leave the area to find other hens or remain in the flock as subordinate males.

Chicken society, therefore, is composed of at least one rooster, his many hens, and their chicks. Roosters are clearly the dominant individuals in the group, lording it over the hens. But the chain of command doesn't end there. The hens themselves each have a position in the group. Hens at the top of the lineup sometimes peck hens beneath them to keep them in their place. This is how the expression "pecking order" originated.

Chickens are not the only animals that organize themselves in dominance *hierarchies*, as these chains of command are called. Chickadees, chimpanzees, gray squirrels, kangaroos, marine iguanas, and, of course, humans are only a few examples of the many species that form hierarchies.

Penguins and ostriches, as we shall see in Chapters 4

and 5, organize themselves differently, but in both cases the males make devoted fathers. In fact, a male ostrich and his primary hen may raise up to 100 chicks at a time, many of them the offspring of other couples!

INSECTS

As varied as avian societies are, however, they don't come close to the wide extremes in social organization seen in insects. In fact, some insects, such as the ants and bees (discussed in Chapter 6), are considered the social experts of the animal kingdom.

The majority of insects, however, live more solitary lives. Among many spider species, for example, interactions between adult males and females are limited to mating. Since most spiders are formidable predators that have no problem eating others of their species, males must approach potential mates very carefully if they wish to avoid becoming a meal. A critical first step in this process is for the male to be sure he's found a mature female of his own species. Usually, males locate receptive females through chemical cues, such as an airborne sexual attractant emitted by the female or the chemicals left on her web.

The second step for the male spider is to make his intentions *very* clear. In some species, males drum on the female's web, while in others they pluck certain strands with their legs. Since these distinctive vibrations are very different from those made by struggling prey, they clearly announce to the female that they are being made by a courting male, not a dinner item. If the female is interested in mating, she signals back by shaking or tapping the web.

In a few species, female spiders feed their offspring until the young are able to trap their own prey. In others, individuals aggregate together and may even cooperate in building a web. However, when most baby spiders emerge from the egg sac, their mother has either died or disappeared and the young are left on their own. With the exception of mating, the rest of their aggressive lives is spent alone.

Like spiders, most butterflies and moths lead a solitary existence. From their birth as caterpillars, through their pupal stage in a cocoon, to their final emergence as winged creatures, these insects have no interaction with others of their kind except to mate. But, in one species, those that are alive at the end of the summer participate in what may be the biggest group event in the animal kingdom—the *migration* of monarch butterflies.

Although the first two broods of monarch butterflies born each year have died by the time autumn comes, the last brood to emerge gathers together in groups and begins to move south from their summer ranges in the northeastern United States, the Great Lakes, and the West Coast to their wintering grounds in Florida and Mexico. These groups are usually spread out rather loosely, but, in years when population levels are high, flocks of migrating monarchs can number in the billions!

Since none of the individuals was alive the previous year, none had ever made the trip before. Yet, each year, they travel along the same pathways their ancestors took—a trip that may be 2,000 miles (3,200 km) for some! Males take the lead, perhaps leaving an odor trail from scent pockets behind their wings for females to follow. How monarch butterflies know where to go is a mystery, but scientists think their behavior is probably programmed genetically.

For many insects, therefore, life—and society—is a pretty solitary affair. However, there are certain insects, such as ants and bees, that depend on each other to a degree that is hard for us to imagine. In their societies, every member's place is determined from birth. Each individual, whether queen, breeding male, or worker, performs a specific, rigidly defined job that is necessary for the survival of the group. All workers, including the soldiers, are female. Males have but one purpose, to reproduce, after which they die. For these animals, living alone is out of the question, since individuals simply can't survive outside the group.

Compared to other animal societies, the societies of these insects often seem more like machines than groups of living animals. Perhaps because these insect societies are so different from our own, they have frequently been used in science fiction as bizarre and frightening examples of alien life.

In contrast, many myths and folktales tell of human children who have been raised by friendly wolves. These stories imply that we think of wolves as similar to us in some important ways. Let's see why this is so.

CHAPTER TWO

The Wolf: Man's Best Friend

Toward the end of the last Ice Age, two very different predators ranged the frozen *tundra* searching for bison, reindeer, caribou, elk, musk-oxen, horses, and other large grazing animals that made these snowy plains their home. Unlike most predators, these two species were capable of capturing animals much bigger than themselves. Equally as important, they were the only predators able to hunt successfully in the wide-open spaces where the large herds grazed—places that provided no opportunity for ambush. Thus, with the exception of a few stray animals snared in rocky gullies by bobcats or lynx, these two groups had the great grazing herds of the Ice Age all to themselves.

One of these predators was the wolf. Although equipped with speed and strong teeth, these things alone would never have been enough to enable individual wolves to capture the much larger and faster herd animals. By working together in *packs*, however, wolves were able to outwit their prey by attacking from different directions, to outdistance them by running in relays, and, finally, to bring them down by fighting cooperatively.

By looking at how wolves hunt today, we can get a good idea of the strategies their Ice Age ancestors used. Unlike large cats, which rely on stealth, wolves make no effort to conceal themselves when they locate their prey. Mingling openly with the herd, they search for an old, sick, or maimed animal, or a young one. Although the wolves are good at locating these weakened individuals, the prey are also good at hiding them in the center of the herd. It usually takes the wolf pack around four hours to finally identify a victim.

Once a suitable animal has been found, the wolves isolate it from the rest of the herd. Then the actual attack begins. One wolf seizes the animal's nose from the front, while a second attacks from the rear. As the prey falls to the ground, the wolves quickly seize it by the throat, opening the jugular vein.

Very large animals are often able to withstand this type of assault, however, and the wolves must try other means of bringing them down. One technique is to grab the victim by the tail, twisting it so that the animal falls. Another is to bite the prey in the leg, severing an Achilles tendon.

In addition to taking individual animals, wolves also trap groups of prey. Several wolves may separate a cluster of animals from the main herd and drive them into an ambush where other wolves are waiting, or they may chase the animals into places where they get wedged in rocks or fall over cliffs. Wolves even drive animals such as caribou or deer into the water. At first glance, this may not seem to make much sense, since wolves are no match for these animals in the water. But the wolves are able to anticipate where the prey will come ashore, and they are there waiting for them.

The second group of Ice Age predators able to overcome the large grazing animals of the tundra was man. While humans did not have the traplike jaws and strong teeth that wolves possessed, they did have the ability to make and use weapons. And, although they had none of the wolves' speed or strength, many of these weapons could be used from a distance, eliminating the need to overtake prey. As in the case of wolves, however, the success of human hunters lay in their ability to work cooperatively. Like wolves, Ice Age humans corraled their prey, herded them into ambushes, and drove them over cliffs.

To hunt cooperatively, both wolves and humans needed a complex social organization and a high degree of intelligence. Not surprisingly, therefore, the two groups developed societies that had—and still have—a close resemblance.

Wolf society is based on the pack, a kind of extended family consisting of a dominant male, a dominant female, their adult offspring, and the dominant pair's young cubs. Although only the dominant male and female mate, all the adults in the pack help to raise the cubs. Like the dominant pair, these "aunts" and "uncles," as the subordinate adults are called, play with the cubs, baby-sit them when the rest of the pack is hunting, and assist in feeding the youngsters and teaching them how to hunt.

A wolf cub's life would seem to be a very happy and secure one. In addition to their littermates, cubs are surrounded by a group of caring, affectionate adults that shower them with attention and are ready to risk their lives protecting them. In this nurturing environment, the cubs form strong emotional bonds to the other members of the pack during the first two months of life.

Since wolves without a territory do not mate, establishing a territory is the first step in forming a pack. Sometimes a male establishes a new territory in a previously unoccupied area or in one in which the resident pack has been exterminated. He then recruits a female from a neighboring pack and the two begin a new pack. More often, however, territories are acquired when one of the dominant pair dies. In this case, the surviving member of the pair does not usually take a new mate, so the territory falls to the next-dominant male. Joined by a female from an adjacent pack, the two start a family and create a new pack.

Most wolf packs have six to ten members, although some may contain up to sixteen or twenty animals. At times, a single wolf will separate from the pack and follow behind it at a distance. Since these lone animals feed on food left over after the pack has finished eating, they are thought to be old wolves whose teeth are worn and who can no longer help the pack in hunting.

Like any group in which the members work together closely and have strong loyalties to one another, wolves must have a well-organized social organization, the ability to recognize each other as individuals, and a means of communicating. Because someone must be in charge in order to lead the others, wolf packs are headed by the most dominant member—usually, but not always, a male.

Competition for the position of dominant, or alpha, wolf begins at a very early age. In playing with littermates, dominant cubs begin to show their leadership qualities, lording it over their more submissive brothers and sisters. Although in some other species, struggles for power between adults may end in the death of the loser, wolves—both adults and cubs—stop the contest when an opponent signals that it accepts defeat.

We are very familiar with many of these submissive signals because we see them used all the time by wolves' domestic descendants, dogs. A wolf or a dog that wishes to say "I give up," rolls over on its back, exposing its vulnerable belly. Other easily recognizable messages of submission include turning the head away so that direct eye contact is avoided, slinking away with the tail between the legs, or groveling close to the ground when approaching a more dominant animal.

Interestingly, humans have many similar body postures that convey the same message. Hanging the head down and avoiding direct eye contact, slinking quietly away, and slumping the shoulders or otherwise assuming a "low profile" clearly signal that we don't want to challenge the other person or cause any trouble—in other words, that we acknowledge that person's dominance over us.

Not surprisingly, the signals used by wolves to communicate challenge are the opposite of their submissive signals. A dominant wolf announces his or her status by direct stares, erect body posture, lifted tail, high head, and a strutting walk. With the exception of tail posture, our body language when we feel confident and assertive is much the same as the wolf's.

Wolves and humans also have similar ways of communicating threat. While you may never have seen a wolf's threatening messages, you probably have no trouble understanding what these same signals mean when sent by a dog. A dog or wolf that feels threatened responds by raising the hair on its back and neck, flattening its ears against its head, narrowing its eyes, snarling and showing its teeth, and emitting a low growl. These messages are intended not only to draw attention to the animal's size and its weapons but also to announce its willingness to fight. Although our hair doesn't stand on end when we are angry, our eyes get narrower, our lips may pull back in a snarl, and our voices often become lower.

By the same token, it is easy to read the friendly greetings sent by wolves and dogs. Even after brief separations, members of a wolf pack communicate their pleasure in being reunited by wagging their tails, licking each others' faces, whining, squeaking, and jumping about playfully—just as domestic dogs greet their owners at the end of the day.

Like humans, wolves also communicate with sounds. In addition to growling, whining, and squeaking, they also howl. Wolves howl to assemble the pack before and after a hunt, to pass on an alarm, to locate each other, and to communicate over long distances. Sometimes, they seem to howl simply for the fun of it. Interestingly, however, they rarely bark.

The communication signals of wolves differ from ours in one important way, however. Humans rarely use smells to convey messages, but odors are a major part of wolf (and dog) communication. Wolves recognize each other primarily on the basis of individual odors, just as dogs identify each other and humans by how they smell. Smells also are very important in mating and in marking the borders of a pack's territory.

Although Ice Age humans and wolves used similar tactics and hunted similar animals, they were not really rivals. Since humans lived in small *nomadic* bands at this time, they were scarcely a threat to wolves. In fact, in some ways, wolves actually benefited from the presence of human hunters. By driving prey into pits and off cliffs, humans often killed far more than they were able to consume. Moreover, with their weaker teeth and jaws, they were able to eat only the most tender parts of the animals they caught. This meant that there was usually a fair amount of food left over from their kills. Since it was certainly easier and safer for wolves to let someone else do the hunting, scientists believe that the animals probably trailed close to human hunting parties, eager to scavenge whatever the humans left behind.

For their part, Ice Age humans were probably not afraid of the wolves following them. According to all available evidence, these northern wolves did not (and still do not) attack humans. Thus, they posed no danger. Although humans may have hunted them occasionally as a source of pelts for clothing, it appears that the relationship between the two was one of peaceful coexistence, much like the relationship between present-day Eskimos and wolves.

With the end of the Ice Age, about 12,000 years ago, the lives of both humans and wolves changed drastically. As the earth warmed, the snowy, open tundras receeded far to the north. In most parts of the world, the huge herds of animals that had fed countless generations of wolves and people disappeared. Thick, dense forests replaced the open spaces, inhabited by smaller and more secretive prey. Both man and wolf had to adjust in order to survive.

Since there were no great migrating herds to follow, groups of human beings began to settle permanently near areas of reliable sources of food, such as a good fishing site or a place with a large supply of edible plants. Wolves continued to hunt, of course, but prey was scarcer and more difficult to find in the deep forests. As a result, they became more and more dependent on what they could scavenge around human settlements.

No one knows exactly when or why people began to bring wolf cubs into their settlements. Perhaps they took pity on orphaned young or enjoyed the antics of cubs at play. But by the end of the Ice Age, if not earlier, wolves and humans had begun living together.

In the well-known children's story *The Jungle Book*, by Rudyard Kipling, a human infant is rescued by a wolf pack and raised along with the wolves' own cubs. As the boy, Mowgli, grows up, he thinks of himself as one of the wolves and considers the alpha male and female to be his parents. Although there has never been a documented case of this happening in reality, the reverse has occurred more times than we can count.

One of the first things humans must have discovered when they took wolf cubs into their homes was that, if adopted during the first two months of life, cubs formed an emotional bond with their human family just like the one they would have formed with members of their own wolf pack. And, not only did they respond to their human owners as if the humans were alpha wolves that should be obeyed, but they also jealously guarded the settlement area as if it were their wolf territory.

Furthermore, since humans and wolves used many of

the same facial expressions and body postures to communicate, it was relatively easy for the wolf cubs and their human masters to understand each other. Thus, because of the close similarities between human and wolf societies, the cubs were ideally suited to become full-fledged members of their new human pack.

As the cubs grew to be adults, humans discovered that guarding the settlement was only one way in which wolves could be used. Even more important were the advantages they offered as hunting partners. Humans locate their prey primarily by vision. This worked well in the open spaces of the tundra, where it was possible to see for great distances, but in the dense forests, vision was not nearly as useful.

In this new environment, their domesticated wolf partners with their keen sense of smell were far more effective. Wolves were also much swifter and able to run down prey that would have outdistanced human hunters. Perhaps most important, however, is that wolves were used to sharing the spoils of the hunt with others of the pack. (Obviously, any animal that expected exclusive rights to whatever it killed would not have made a very good hunting partner from a human point of view!)

These early wolf partners clearly did not stay wolves, however. Anyone familiar with dogs knows that most breeds are very different from their wolf ancestors. When did wolves become dogs, and how did this happen?

The transformation probably started as soon as people began raising wolf cubs. A trait that was critically important to humans was how well the cubs accepted people as leaders of the pack. Various cubs must have differed in this regard, with some maturing into adults that actually challenged humans for the dominant position. Those animals that refused to consider themselves subordinate to humans were probably killed or at least thrown out of the settlement, as they would have been useless, if not dangerous, to have around.

Thus, only the more submissive animals were allowed to stay. As these animals reproduced, they passed on this trait of submissiveness to more and more of their offspring, with humans continually eliminating the few animals that showed unwanted aggressiveness. The process of selecting certain traits and allowing only individuals that have those traits to breed is called *selective breeding.* This technique was used to mold wolves into animals—now called dogs—that more closely fit human needs.

One alteration humans made was to create animals that barked more, so that hunters could keep track of the faster-moving dogs as they followed prey. Barking was also a way that dogs could alert humans to danger from a distance or during the night when their masters were sleeping. Similarly, people took the wolves' ability to herd prey into ambush and selected for breeding those wolves that would herd domestic animals without subsequently killing them.

All of the traits that various breeds are famous for, as well as the many different shapes of their bodies, are the result of human selective breeding. But people altered wolves' social nature very little in this process. They didn't have to—in all the important ways, the wolves' society was close enough to ours in the first place.

CHAPTER THREE

Not As Unfriendly As They Seem: Alligators, Sharks, and Rats

Alligators, sharks, and rats have a decidedly poor public image. Viewed by people as machinelike killers, these animals show a different side to others of their own species. In several respects, their societies are far more advanced than most of us suspect. But because we do not understand their ways of communicating, we frequently misinterpret their behavior, sometimes with disastrous results. As we have learned more about these animals, however, we've come to realize that they are not quite as unfriendly as they seem.

By most standards, alligators and crocodiles can hardly be considered very social animals. Both tend to be territorial, with individuals defending their basking sites and the surrounding area and excluding others from the vicinity. During the breeding season, aggression between males escalates dramatically, with individuals often viciously attacking each other. At night, whole swamps echo with their bellowing roars as they attempt to attract females while simultaneously driving off rival males.

Although adult alligators and crocodiles have few enemies, the same cannot be said for hatchlings and juveniles. Any number of larger mammals, birds, and other reptiles consider them fair game, and mortality among the young is very high. In this inhospitable environment, newly hatched alligators and crocodiles need all the help they can get. Fortunately, their mothers are willing to provide it.

A few snakes and lizards guard or tend their eggs, but alligators and crocodiles are the only reptiles that actually care for their young after they are born. In both these groups, maternal concern begins with the construction of a special nest for the eggs.

In the case of alligators, this nest is a large mound made of plant material and mud that measures over 7 feet (2 m) in diameter and 3 feet (almost 1 m) in height. When she has finished piling the plants and mud into a heap, the female climbs to the top and digs out a hole in the center. Carrying still more mud and plants from the shore in her mouth, she fills the hole with a mud/plant plug that serves as a receptacle for her eggs. After laying the eggs, the female alligator covers the top with another load of vegetation and mud and then packs the mound down firmly with her body until it is smooth and cone-shaped.

The heat generated by the decaying plant material makes the nest a perfect incubator for eggs. The huge mound is very conspicuous, however, and if left alone, there would be many animals eager to dig the eggs out and eat them.

The female doesn't give them the opportunity, however. She patrols the nest area constantly, sometimes sprawling on top of the mound. Needless to say, her mere presence is sufficient to ward off any potential predators.

Guarding the eggs is not her only reason for remaining near the mound, however. When the young hatch, they will need their mother's help if they are ever going to get out of the nest.

Deep within the mound, the baby alligators make a high-pitched squeaking sound when they hatch that lets their mother know it is time to dig them out. Carefully, the female pushes the packed vegetation away with her claws, freeing the hatchlings. Then, one at a time and *very* gently, she takes them into her mouth and carries them to the water.

Once in the water, the babies stay close to their mother, protected by her for a year or more. If threatened, they make a special distress cry, alerting her to the danger. The female responds by racing over and taking the youngsters into her mouth—undoubtedly the safest place on earth for a little alligator. Ironically, people used to think that the females were eating the young when they did this.

The behavior of female Nile crocodiles is very similar to that of female alligators. Since they live in a much hotter climate, however, their nests are not large mounds intended to generate heat but shady sand pits that will keep their eggs cool. Like her American relative, the Nile crocodile remains near the nest, both to protect the eggs and to dig the hatchlings out. These females also escort their babies to the water and remain with them while they are young.

SHARKS

There is much that sharks have in common with alligators and crocodiles. Among the very oldest creatures on earth, they all seem to have a deeply disturbing, primitively savage quality about them. Their ability to sever limbs with a single snap of the jaws or to devour an entire human being in minutes strikes terror in our hearts.

Although this is the image most of us have of sharks, it is not how they are viewed by many people around the world—people who have grown up around these ancient fishes and who understand their behavior. Far from being motivated by a voracious hunger or the sheer desire to kill, sharks are generally shy creatures that are not usually interested in us as food. Most often, it is our own actions that cause them to attack.

While sharks do not usually eat people, there are several situations that can lead them to mistake us for their natural food. Sharks are attracted to blood or other signs that a fish is wounded and thus easy prey. Therefore, cuts on a swimmer or speared fish tied around a person's waist can easily make a human look like shark food. Thrashing about in the water sends signals that resemble those of wounded fish, and metal objects worn by swimmers can create electric currents like those of vulnerable prey. For all these reasons, it is easy to see why victims of an accident at sea are in such danger.

Most shark attacks, however, occur because a swimmer has disturbed the fish or invaded its territory. Sharks have distinct signals that warn other sharks—and people—that they are trespassing and must leave if they don't want to be attacked. By turning and twisting its body, shaking its head, humping its back, and dropping its pectoral fins, a shark clearly announces that it is feeling threatened and is getting ready to defend itself or its territory. Unfortunately, most swimmers don't understand what the fish is trying to say and thus don't heed the warning.

RATS

Although tiny in comparison to alligators or sharks, rats have a similar reputation for vicious, unbridled aggression.

Portrayed as uncontrollable hoards swarming over anything in their path, rats are believed to be inherently mean, fighting among themselves and frequently killing each other in the process. Far from being protected by adults, the young are thought to be in particular danger from them, with males ready to annihilate newborns at the first opportunity. Even the well-known scientist Konrad Lorenz compared them to humans in terms of their readiness to murder their own kind.

The truth is, however, that rats live in well-ordered, peaceful colonies where strangers, if they behave correctly, are tolerated. Serious fights among individuals are rare, and adult males do not pose a special threat to infants. Unlike "higher" primates, male rats do not use their superior strength to monopolize group resources. In fact, not only are rats especially well suited to group living, but their social organization has been a major part of their success in competing with people.

Rats communicate through odors and high-pitched sounds that are out of our hearing range. Our inability to tell when they are communicating with each other, much less what they are communicating about, is part of the reason we have had difficulty understanding life in rat society. With the use of special equipment, however, scientists can now detect these messages. This has helped them understand how most rats manage to avoid attack when they enter a colony of strangers.

Like many other animals, rats have ways of signaling submission that ward off aggression from other individuals. In their case, the signal consists of special ultrasonic sounds. An outsider rat that enters a colony is usually accepted by the resident rats if it makes these sounds. If the rat fails to send the correct submissive signal, however, it is attacked.

Thus, instead of being helpless victims, individual rats can generally control whether they will be attacked or not. Most animals, of course, prefer to avoid a fight and send the appropriate message. However, a dominant rat used to winning aggressive encounters may decide not to submit. Although rats that do this can expect to be attacked, their experience has taught them that—for them—fighting pays off.

The myth that adult males readily kill infant or juvenile rats is also untrue. In fact, experiments show that rat pups raised in the presence of both their father and mother do better than those raised by their mother alone. When the mother is removed, the father usually behaves quite maternally toward the youngsters. In the wild, millions upon millions of rat pups have been successfully raised in mixed colonies of males and females. Obviously, if male rats were as murderous as reputed, the species would have died out long ago.

In the war waged on rats by humans, poison is a major weapon. Although the common view of rats is that they gorge voraciously on anything they can, in reality these animals are very restrained in their eating behavior and extremely careful about their choice of food. In fact, their habit of avoiding unfamiliar food, eating in groups, and consuming only what other rats safely eat makes it difficult to poison them. If some rats become sick, the type of food they ate is avoided by others in the colony. Thus, rats' ability to learn from each other is one of the main reasons why they are still around.

Finally, in terms of equality between the sexes, rat society appears to be a feminist dream. In most of the mammalian and avian societies in which males and females live together, males are dominant over females. Among other things, the dominant position of males guarantees them first access to food, even when females have found or caught it. Thus, when food supplies are limited, it is the females that have less to eat.

Interestingly, this does not happen in rat society. Although male rats are bigger and stronger than females, their physical dominance does not grant them first place at the table. Instead, males and females share equally in the colony's food resources. Successfully raising young in most species is dependent upon females having enough to eat, especially in the case of mammals that must nurse their newborns. Thus, the willingness of males to share food equally with females is one reason why rats have been so successful in increasing their numbers.

CHAPTER FOUR

Watery Societies: Dolphins and Penguins

In general, the social interactions of most fishes are limited to schooling and mating. Many mammals and birds that have taken to the sea, however, have developed complex societies. The most famous of these, and the most evolved, is that of dolphins.

DOLPHINS

A number of scientists believe that dolphins may be almost as intelligent as humans. As evidence for this, they point to the highly developed social nature of these animals, their skill in using items in their environment as tools, and—most important—their apparent ability to learn simple languages taught to them by researchers.

Dolphins have their own ways of communicating as well, but since it is impossible to follow free-ranging groups in the ocean, recording all the sounds they make, we do not completely understand how this signaling system works. An additional obstacle in learning about dolphin communication is that, as with other aspects of their lives, dolphin signals are designed for a watery environment. Since the sounds they make are four and a half to ten times higher in pitch than our vocal sounds, many are out of our hearing range. Thus, without special equipment, we miss much of what dolphins say to each other.

Our voices, on the other hand, are at the lower end of dolphins' hearing range. In fact, not only can dolphins hear us talk, but their hearing is sharp enough to pick up sounds in our speech that we can't hear ourselves. It is no surprise that dolphins' hearing is far more acute than ours, or that their eyesight isn't as good. In water, hearing is of much greater use than vision.

Although dolphins may look like fish, their behavior and social organization are distinctly mammalian. In fact, except for differences in *habitat*, life in dolphin society is very similar to that in the elephant community described in Chapter 1.

Like most mammalian societies, dolphin communities are centered around mothers and their young. Dolphin females make extremely caring parents, and the death of a baby is visibly upsetting to its mother. Since they are mammals and not fish, dolphins breathe the same way we do and must come to the surface regularly for air. There are many reports of females supporting their dead infants at the surface, sometimes for days, in an effort to revive them.

Baby dolphins are born tailfirst and swim to the surface by themselves for their first breath. The mother then guides the infant to a position just below her dorsal fin. Here it is carried effortlessly along on the pressure waves created as the mother moves through water, similar to the way a race car driver "drafts" on the car in front of him.

Young dolphins may nurse for years, even after they have learned to eat fish. On the average, a youngster, or calf, spends about five and a half years with its mother, swimming with her constantly until the birth of her next calf.

Like young elephants, little dolphins are raised in a group composed of several older females, their female descendants, and all their dependent offspring. Periodically during the day, this group breaks up into smaller bands of mother and offspring pairs, with mothers that have youngsters of the same age tending to stay together. This means that juvenile dolphins begin to develop relationships with others of their own age from the day they are born.

As they grow older, adolescents separate from their mothers and form independent groups. It is usually these animals that we see cavorting playfully with each other in the waves. During this time, each male forms a long-lasting friendship with another male, the two often having grown up in the same smaller band. Members of these pairs are inseparable, with some of these friendships lasting ten to fifteen years. After reaching sexual maturity, the pair travel together from one group of females to the next, seeking mates.

Dolphins maintain relationships with their relatives for most of their lives. When females have their first baby, many return to their mother's group to live, and both males and females often come back for a visit when their mother gives birth to a new infant.

Their close personal relationships and their ability to work together as a group are largely responsible for dolphins' survival. This is particularly true in defending themselves against predators such as sharks. Interestingly, dolphins are far from helpless when faced with shark attacks. Many sharks avoid schools of dolphins, and for good reason. By ramming a shark's vulnerable gills or belly with their snouts, dolphins can easily kill it.

Just as mothers often lift their dying or dead infants to the surface to breathe, adults support other adults that are sick or injured so that they don't drown. The helpers continue to carry their disabled companion at the surface until it recovers or dies. Dolphins have even been known to aid drowning humans in this way, holding them afloat and pushing them toward shore.

Cooperating as a group also has advantages in obtaining food. Although a dolphin catches most of its meals by itself, it occasionally works together with other dolphins to herd schools of fish into a small area. Then the dolphins take turns swimming through the trapped fish to feed.

Young dolphins have a lot to learn before they become independent, and living in a group is one of the best ways to acquire the knowledge they need. By observing others, they learn to recognize prey and how to catch it, to identify predators and defend themselves, to define the boundaries of their home range, and to recognize other dolphins and determine their place in the group hierarchy.

Perhaps the aspect of dolphin society that has most interested scientists, however, is the way in which these animals communicate with each other. Dolphins signal each other through the use of a variety of high-pitched clicks, whistles, quacks, barks, and growls. Like many other animals, they use these sounds to locate each other, court mates, alert each other to danger, establish dominance, and identify themselves.

Baby dolphins can whistle from birth. Before they are a year old, they have developed their own special signature whistles that identify them as individuals. Although male youngsters develop signature whistles that are similar to their mothers', females create ones that are distinctly different. This is because most females continue living with their mothers when they grow up. It is important that the whistles of daughters and mothers differ if others in the group are going to be able to tell them apart.

Males go off on their own when they mature, so there is little danger that their whistles will be confused with their mothers' signatures. Since males often form coalitions with other males, some scientists have suggested that having

With ears spread and trunk raised, a matriarch prepares to defend the baby elephants hiding underneath.

If this chimp is a dominant female in the troop, her baby is likely to grow up to be a high-ranking adult, too.

Unlike most frogs, this female arrow poison frog protects her two tadpoles by carrying them on her back.

By displaying the brightly colored skin on his throat, this male anole lizard tells a rival to get out of his territory.

Above: The gopher snake lays her eggs, then leaves them to hatch on their own.

Facing page: Because all newborn mammals nurse, they must be cared for by their mothers when young.

Unlike many baby birds, ducklings can swim and feed themselves from birth.

It takes both parents to feed this totally helpless baby oriole.

Both blue-footed booby parents take turns
shielding their egg from the hot equatorial sun.

Above: By plucking certain strands or drumming on her web, male spiders ask females to mate.

Facing page: Male bears spend most of their lives alone.

Migrating monarch butterflies cover whole trees when they stop to rest.

Even working together, these wolves will have a hard time taking the deer carcass away from the bear.

Although most newborn reptiles must fend
for themselves, alligator hatchlings are
lucky—their mother is *very* protective!

By living together in groups, dolphins can more easily defend themselves against sharks, capture prey, and teach their young the skills they will need to survive.

Adélie penguins perform a special ritual
when they change guard duty at the nest.

Above: A male ostrich, with his black and white feathers, takes off after the female on the right, while another female on his left tries unsuccessfully to catch his eye with a mating display.

Below: Using their bodies as a bridge, weaver ants join leaves together to make a nest.

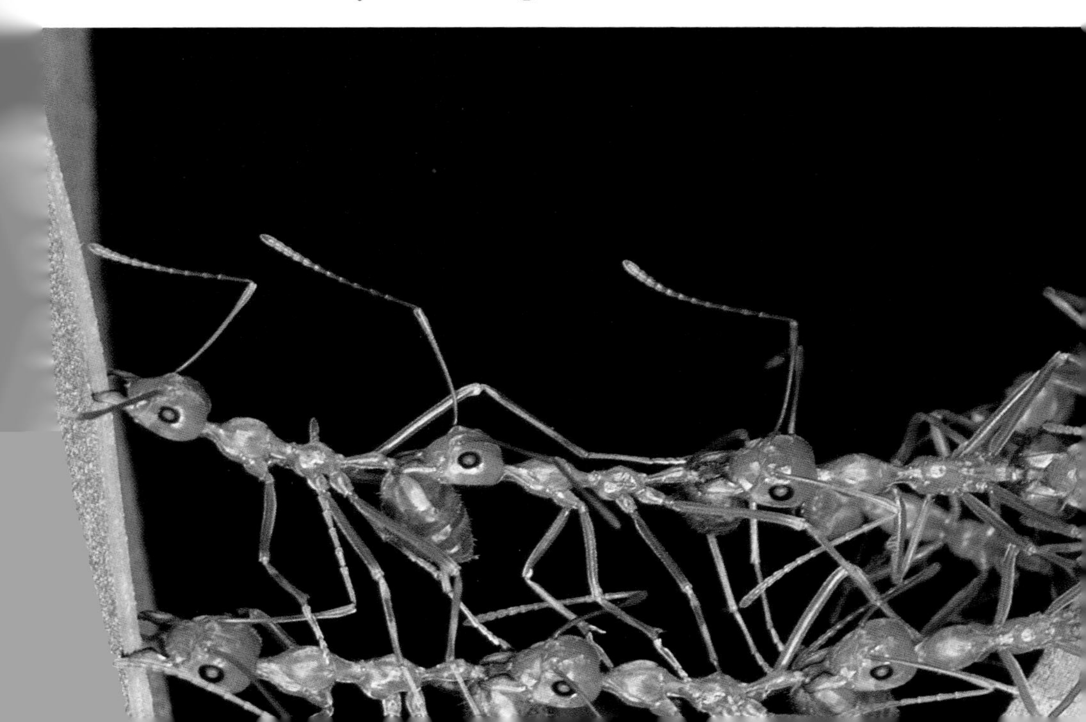

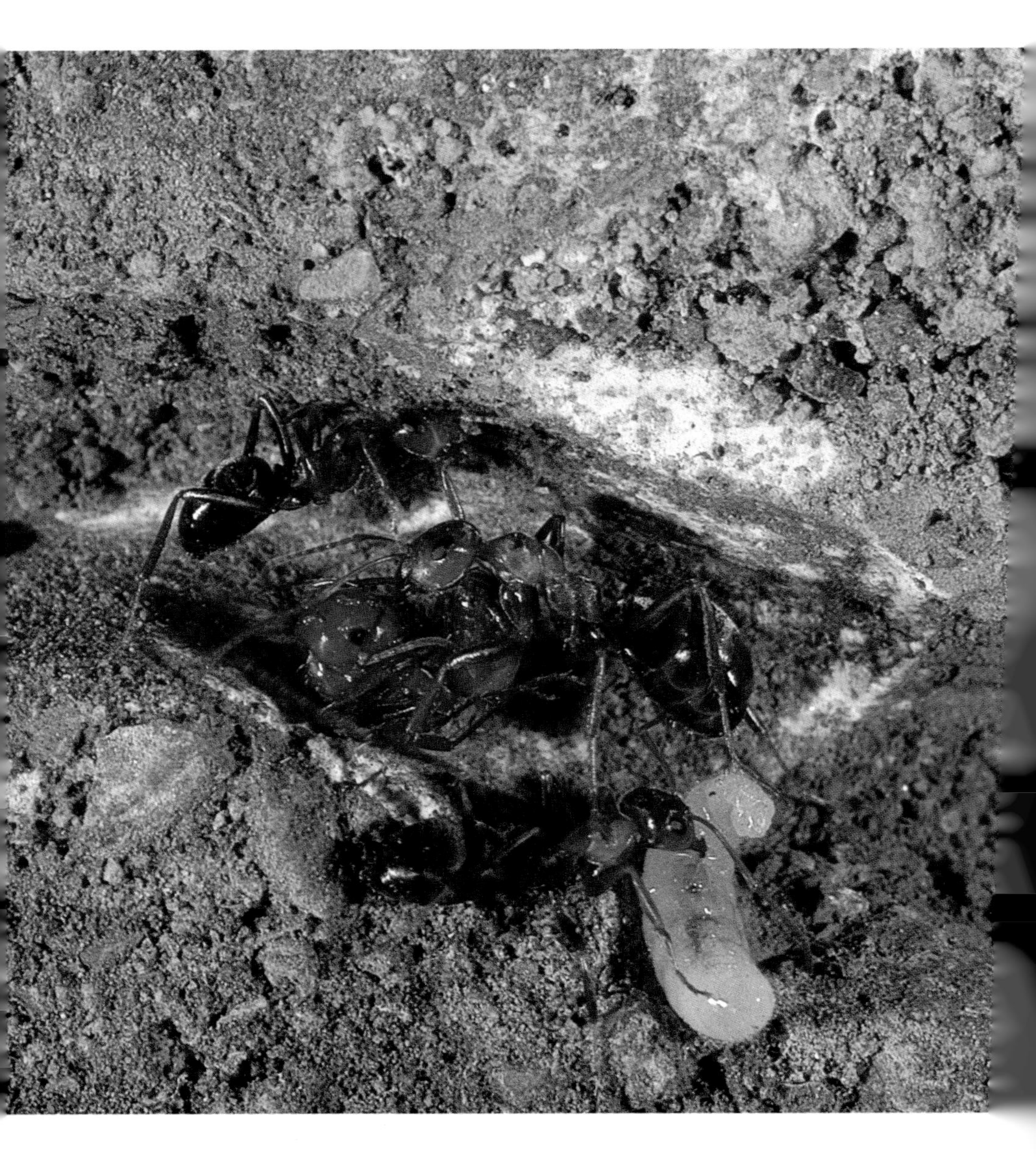

Formica slaves (black) do all the work
for their Polyergus (red) colony mates.

whistles that resemble their mother's may allow brothers to recognize and team up with each other. The fact that sons and mothers have similar whistles may also help them to avoid in-breeding.

It appears that dolphins may sometimes emit *another* animal's signature whistle in order to call that individual, just as we call someone by name when we want that person in particular to respond. Although this may seem unbelievable, it is not any more amazing than the ability of some dolphins to learn special languages taught to them by researchers. Dolphins have shown that they can learn a sound vocabulary and a sign language vocabulary, as well as the grammatical rules for putting these artificial words together in sentences. People who work with dolphins have joked that perhaps they are *more* intelligent than we are. After all, they have been able to learn some of our languages, while we don't have a clue as to what most of their signals mean.

PENGUINS

In many respects, the adjustments penguins have made to living in a watery environment are as great as those of dolphins. While dolphins are mammals that can no longer come ashore, penguins are birds that can swim but can no longer fly. Just as dolphin society is recognizably mammalian, penguin social organization is typical of that of many birds.

Although their comical forms are familiar to most of us, we are often not aware of how radically evolution has changed the bodies of penguins. Not only do these birds spend much of their lives in the water, but it is the coldest water in the world. Several physical alterations were necessary if penguins were to survive in such an inhospitable environment.

The lightweight bodies and hollow bones of most birds are what make it possible for them to fly. Penguins dive deep into the ocean for their food, however. If they were built like most airborne birds, they would be unable to stay underwater for long. Without more weight, their bodies would quickly pop to the surface like corks.

Over time, penguins developed the heavy body and solid bones that provide the ballast necessary to weigh them down. Other changes occurred as well. Their wings became modified into powerful flippers, enabling them to swim underwater at speeds of over 5 miles (8 km) per hour. Penguins also developed a thick layer of body fat and short, dense feathers to help keep them warm.

The end product of these evolutionary changes was a bird that is too heavy to fly but is a champion diver and swimmer. Not only are penguins extremely fast underwater, but they can dive for several hundred feet, remaining there for two to three minutes at a time.

There are seventeen species of penguins, each of which has its own breeding cycle, mating behavior, and nesting habitat. Some groups nest in burrows, while others build nests aboveground. The king penguin doesn't build a nest at all, but balances its one egg on the top of its feet! Penguin colonies also vary in size, from just a few nests to more than a million. And, although most species inhabit the Antarctic or nearby areas, one little penguin lives on the Galápagos Islands, right on the equator. In spite of these differences, however, penguin societies have much in common.

Adélie penguins provide a good example of what life is like in a large penguin colony. As with most birds with helpless young, both parents must help in raising the babies if they are going to survive. Adélies, like other penguins, are monogamous, with males and females having only one mate at a time.

In fact, in Adélie society, many individuals mate for life, returning to the same partner year after year. Since raising young in the harsh Antarctic requires flawless cooperation between parents, an individual has a better chance of success if it chooses a partner with whom it has worked well in the past. On the other hand, couples that have been unsuccessful in reproducing have nothing to gain by staying together and tend to pick new partners the following year.

Like most other penguins, Adélies spend the winter feeding in the ocean. As summer approaches (the beginning of October in the southern hemisphere), the sea ice begins to break up and the birds begin to come ashore for the breeding season. In a way that scientists don't completely understand, each penguin finds its way back to the place where it was born or, if it is older, to the area it nested in the year before.

Male Adélies generally arrive first and begin immediately to stake out a territory. Although the territories of many other birds can be quite large, penguin territories consist of only a nest and the small area around it. One reason for this is that penguin eggs can't be left unattended, even for a few minutes. A penguin has to be able to defend its turf while still sitting on the nest. This limits territory size to a small circle of 2 to 2½ feet (0.6–0.8 m) in diameter.

As soon as he has claimed an area, the male starts building a nest from stones and small pieces of old bone. Building materials in this environment are in short supply, so individuals steal pebbles from each other's nests constantly.

While the early arrivals have been establishing territories and building nests, females and straggling males have been arriving. Pairs are reunited, recognizing each other by individual differences in their cries. If last year's mate does not turn up, the remaining bird picks a new partner. Sometimes, a previous mate arrives after its partner has chosen another. This embarrassing circumstance often results in a violent squabble, with the bird in the middle sometimes returning to its old mate, at other times staying with the new one.

In early November the female lays two eggs. Now the stakes become much higher. With predators such as the gull-like skuas hovering overhead, ready to swoop down and snatch the eggs, any lapse on the part of the nesting couple can be fatal.

By this time the female, which has not eaten in almost a month, takes off for the ocean to feed. The male, which has been fasting at least this long, remains on the nest to guard the eggs. The female may be gone anywhere from a few days to three weeks. When she finally returns to take her shift incubating the eggs, the poor male may have lost 40 percent of his body weight!

Whenever there is a changing of the guard at the nest, the couple perform a special nest-relief ceremony. Both the male and female stretch their necks up, pointing their beaks into the air and waving their heads back and forth while they let out raucous cries. Then together they point their beaks at the ground, open their eyes wide, and growl. As the growling grows louder, the birds raise their heads and open their beaks wide. When their beaks are pointing straight up, the couple take a momentary break, then start all over. Finally, the returning bird steps behind its mate and onto the nest.

Strange as this practice may seem, it is a very important ritual in Adélie society, as are similar ceremonies among other penguin species. This nest-relief display probably serves several purposes. For penguins, aggressiveness during the breeding season is a very advantageous trait to have. A breeding couple must defend itself constantly, not only from predators that wish to steal the pair's eggs, but from neighbors that are busy trying to grab nest material from under their very noses.

The one individual a bird does not wish to be aggressive toward, however, is its mate. The nest-relief ritual probably helps to reduce aggression between mates while at the same time giving the couple an opportunity to identify each other. Interestingly, a returning bird may also present its mate with a gift of a small stone. Bringing a stone, instead of stealing one, may be a way of signaling friendly intentions, thereby also reducing tensions between the pair.

In December the eggs hatch. For the first ten days or so, the nestlings are so vulnerable to cold that they will freeze to death if left uncovered for even a few minutes. Therefore, the parents must again take turns providing constant shelter for them. While the young huddle under one parent, the other goes to sea to bring back food for their very hungry mouths.

When the nestlings become too big to fit under one parent, they graduate to a kind of penguin kindergarten, or *crèche*, where they cluster together with other youngsters their age. Although the adults are no longer guarding the chicks, skuas do not seem to attack them when they are grouped together. Huddling together in the crèche also helps keep the chicks warm during a storm.

Although the chicks can now withstand the cold air, they do not yet have the waterproof feathers which permit penguins to survive in the sea. Thus, their overworked parents must still feed them.

It is fortunate that the chicks no longer need their parents for shelter, because with the youngsters' increasingly voracious appetite, it takes the efforts of both adults to bring in enough food. The young demand a meal from any adult coming out of the water, but parents feed only their own offspring, recognizing them by their individual cries.

Finally, as the brief summer draws to a close, the youngsters acquire waterproof feathers and are able to swim and feed themselves. By March they are completely on their own. With the departure of the birds soon after, the biggest social event in penguin society—the breeding season—is officially over.

CHAPTER FIVE

The Altruistic Ostrich

Penguins are not the only birds that can't fly, of course. Ostriches and their smaller relatives, the rheas of South America, are also earthbound. Weighing between 200 and 300 pounds (90–135 kg), it's no wonder ostriches can't get off the ground. Their size is not the only thing that makes these birds unique, however. Not only are ostriches the world's largest bird, but their society is unlike that of any other animal in the world.

In the vast majority of animals, the young are able to care for themselves or are raised by their parents. In some species, such as the wolves (Chapter 2) or the social insects (Chapter 6), adults may raise young that are not their own offspring, but when this occurs, the young and their caretakers are always relatives.

Although a few animals are known to bring up young they are not related to, they must be tricked into doing so. For example, farmers who wish to persuade a female sheep to accept an orphan lamb must cover the youngster with the hide of the female's dead offspring. Only by convincing her that she is feeding her own baby will the mother care for the orphan. Cuckoos lay their eggs in the nests of other birds, leaving them to do the work of incubating the eggs and feeding the baby cuckoos. The foster parents do this because they are fooled by the close resemblance between the cuckoo eggs and their own.

The willingness of most animals to raise only relatives makes evolutionary sense. Freeloaders (those that leave the raising of their offspring to others) have the time and energy to produce many young because they don't have to take care of them. Since they don't feed their babies, they don't have to share the food they collect. By leaving the protection of their offspring to others, they also reduce the times when they expose themselves to danger. Thus, they live for more breeding seasons and produce more offspring per season.

In comparison, altruistic animals (those that are willing to raise unrelated offspring) are at a great disadvantage. Not only does the stress of raising and protecting young take a toll on them, reducing their life span, but only some of the young raised are even their own. If animals were willing to raise nonrelatives, freeloaders would leave a great many more descendants than would altruistic individuals. Eventually, the freeloaders would outbreed the altruistic individuals. Soon, there would be no one but freeloaders left, and the species would die out.

Ostriches are a perplexing exception to this rule. Not only are adults altruistic, raising young they are not related to, but they actually compete for the privilege of doing so. Yet ostriches actually profit from this strange behavior. How can this be?

Part of the reason stems from the fact that female ostriches outnumber males. Females have drab feathers that camouflage them, while the black-and-white plumage of male ostriches is much more dramatic. Unfortunately, while their coloration is very important in helping males attract a mate, it also makes them more visible to the many predators of the African *savanna* where they live. By the time ostriches mature, there is only one male for every three or four females.

Other things make the supply of breeding females even more lopsided. Females become sexually mature at an earlier age than males. In addition, only males with territories can attract a mate. Therefore, some of the available males are not acceptable suitors.

The cooperation of both parents is needed to protect the eggs and the young in this dangerous environment, so, unlike female bears, female ostriches cannot raise their babies by themselves. With not enough males to go around, it would seem that some females must be left out. Fortunately, ostriches have discovered an interesting solution to this dilemma that lets all females breed.

The ostrich breeding season begins with the establishment of territories by adult males. As mentioned, not every male is able to carve out a space of his own; in fact, only about half the males are successful. Approximately a month later, females visit the territories of neighboring males, looking for a mate. About one in three females finds a permanent mate and lays her eggs in the nest he has prepared. These females are called major hens, and they form a bond with their mates that can last several years.

The remaining hens mate with several territorial males but do not form bonds with any. Since they do not have nest sites of their own, they must lay their eggs in the nest of a major hen. It is easy to see why minor hens are willing to do this. Since the male and his primary mate will take care of the eggs (and later the chicks), the minor females can enjoy the advantages of reproducing, with none of the risks. In a sense, minor hens freeload on the care and protection offered by the male and his primary mate.

The puzzle is why the male and his major hen permit them to do this. Of approximately forty eggs in the nest, only about seven belong to the couple. Presumably, few if any of the rest are from relatives of either the male or the major hen. Some of the eggs may be the result of matings between the male and minor hens, but certainly not all. Yet, the male and his primary mate incubate all of the eggs.

Like penguins, ostriches take turn incubating the eggs and guarding them. Vultures, hyenas, jackals, and lions all await the chance to swoop upon an unprotected clutch and eat the eggs. Even the adult ostriches themselves are in danger of attack and are sometimes killed while defending the nest.

The couple's job is not finished when the eggs hatch. In fact, it seems that the male and his major hen actually ask for more work. Young chicks must be constantly protected against the same predators that threaten eggs. Although ostriches are capable of killing even a lion with a well-placed kick, they lose many battles as well. It is then difficult to understand why a dominant male and his major hen try to round up the young of other couples to join their own flock. A pair that is successful in this competition can end up sheparding up to 100 chicks, all of which must be defended. Why does the couple risk their lives for the offspring of strangers?

At this rate, it would seem that freeloading ostriches will outbreed altruistic pairs in record time, but surprisingly, that doesn't happen. A closer look at ostrich society indicates that the behavior of a male and his major hen is not as self-sacrificing as it might seem.

Scientists realized that there might be more going on than meets the eye when they looked at the way eggs are placed in the nest. Although a female's body is large enough to cover only twenty to twenty-five eggs, approximately forty are laid in the nest. Obviously, about fifteen to twenty of these eggs have to go. The major hen keeps her own seven eggs in the center of the nest and tosses some of the excess eggs out. Strewn around the rim of the nest, the ejected eggs are the first to be taken by predators. Even if eggs from the nest are stolen, the major hens's eggs are at the center and thus are less likely to be victimized. By contrast, if the major hen allowed only her own eggs to be laid in the nest, every egg taken by a predator would be one of hers.

Dominant pairs have a similar reason for wanting to recruit other chicks to their flock. By grouping their own offspring with others of the same age, the couple reduces the chance that theirs will be killed by predators. Seen in this light, the male and his major hen can hardly be called altruistic.

This arrangement may appear to be a bad deal for minor hens, but it actually has some advantages for them. Although their eggs and chicks are more likely to be taken by predators, at least some of them survive. Without a territory of their own, minor hens would not be able to raise any chicks at all, so they benefit by laying their eggs in another's nest.

While ostrich society differs from others in appearance, it is not really so strange after all. Like the societies of other animals, it serves one purpose: to help individuals of a species survive and reproduce in the particular environment in which they live.

CHAPTER SIX

The Social Experts: Ants and Bees

In all other animal societies, when an individual is unable to reproduce, it is usually because it cannot find a mate. There are many reasons why it may be difficult to attract a partner. For example, an animal may have been unable to establish a necessary territory or it may hold a low-ranking position in a group where only the dominant individuals breed. With the exception of a few animals that are unable to breed because they have a physical defect, however, any individual can have young if given the chance.

Among ants and social bees, however, reproduction is not an individual but a group affair. In fact, the vast majority of individuals in these societies are physically unable to breed at all. As a result, the entire group devotes its efforts to raising the young of only one female. In the case of ants, this can mean that hundreds of thousands of colony members care for the offspring of a single female.

Ant colonies, like the hives of social bees, are composed of a queen, workers, and breeding males. The sole function of the males is to fertilize a queen, and after mating they die. As the reproductive female of the group, the queen spends her life churning out eggs. All other tasks necessary for the survival of the community, including raising the young, are performed by the workers, a huge class of sterile females. Since these workers are all offspring of the queen, they are all sisters.

Ant and bee queens are dramatic proof of the saying, "You are what you eat." All female larvae have the potential to become reproductive females, but whether they will or not depends upon the type of food they are given. Those that will grow into queens are fed special secretions by the workers that raise them.

Since ants and social bees are unable to live or reproduce on their own, the welfare of the colony or hive is critical to their survival. As a result, these insects have developed group cooperation into a fine art. Weaver ants are an excellent example of what some of these insects can achieve by working together.

ANTS: WEAVERS, STONE-THROWERS, AND SLAVE-MAKERS

The weaver ants of Africa and Asia are extremely aggressive and territorial, with colonies consisting of as many as 500,000 workers. The large "major" workers are general laborers, responsible for foraging, nest construction, feeding the queen, and fighting the colony's enemies. Within the nest, smaller "minor" workers care for the queen's eggs and feed and wash the tiny larvae that hatch from them.

In this warlike insect society, the territories of neighboring colonies are usually separated by a kind of demilitarized zone where few ants from either group venture. In spite of this "no ant" zone, individuals of different colonies sometimes bump into each other, triggering a full-scale battle between the two groups.

Soon, masses of ants are engaged in the fight. Snapping viciously at each other with their mandibles, the combatants pin their enemies to the ground, clipping off their legs and antennae, and ripping open their abdomens. Warfare between two colonies can go on for days. Finally, one colony succeeds in defeating the other, and the vanquished ants retreat from the territory.

Although their military abilities are certainly impressive, weaver ants are even better known for the way in which they work together to build, or "weave," their nests. Once they have chosen a suitable tree branch, workers begin to try to fold over the tips and the edges of the leaves. As soon as one ant succeeds in pulling back a leaf tip, the others abandon their efforts and join her. When the leaf is broader than the length of an ant's body or when the ants are trying to pull together two leaves separated by a wide space, the workers form a living bridge across the gap. Some of the ants in the chain then crawl on top of the others and pull backward. This shortens the chain and brings the leaf edges together.

When the leaves are folded back sufficiently, some of the workers hold the edges down with their legs and mandibles, while others race back to an established nest for something to tie everything permanently in place. The workers return with partly grown larvae, which they wave back and forth across the leaf seams. This stimulates the larvae to release threads of silk from glands below their mouth. Using thousands of these strands, the workers secure the leaves in position and a new nest is created.

Weaver ants are also noted for one of the ways in which they acquire food. Although many ants prey on insects, weaver ants are among the few that maintain insect "herds." Colony members guard certain sap-feeding insects as if they were dairy cattle, collecting their sweet-tasting excrement for food.

The complicated activities of weaver ants—colony defense, nest-building, exploration, emigration to a new nest site, foraging, hunting, herding, caring for the queen, and raising young—could not be carried out unless these insects had a system for coordinating the actions of individual colony members. Like other ants, weavers use a variety of chemical and tactile (touch) signals to communicate. By combining these signals in different ways, the ants can send complex messages telling each other when they should act and what to do.

In addition to weaver ants, several other species also have unique ways of cooperating to get food. The leaf cutter ants, for example, do not eat the leaves they collect. Instead, they use them to make gardens where they grow their special food, a fungus.

Other ants cut down on the competition for food by stoning their competitors. A little Conomyrma ant of the American Southwest has found an ingenious way to keep a neighboring species from foraging. The ants pick up small pebbles in their mandibles, carry them to the rim of their neighbors' nest entrance, and drop them inside. Although ten to thirty workers can be involved in a stone-throwing seige, fewer than five workers are enough to keep their targets penned inside the nest throughout the night (when both species forage). Sometimes victims of this technique are unable to forage for several weeks.

Scientists are not sure exactly why stone-throwing works, but it is evidently not because the stones hurt the target ants or seal the nest entrance. Slow-motion cinematography of stone-dropping attacks shows that very few victims are actually hit. When researchers toss similar stones down the nest entrance, the ants simply remove them. Even when hit directly by the stones, the ants go about their business as usual, seemingly unaffected.

Scientists think stone-dropping may serve as some kind of signal, one that has a meaning only when it is sent by another ant. If this is so, it is the only instance we know of in which ants have used an object from the environment, rather than something produced by their own bodies, to communicate.

Perhaps the most interesting of all ant societies are those of the slave-making ants. Thirty-five ant species depend to some extent on slave labor for their existence. Certain groups, such as the Polyergus ants, could not survive at all without their slaves.

Polyergus workers are incapable of foraging for food, feeding their young, caring for their queen, or even maintaining their own nests. However, they are very good at one activity—raiding the nests of other ants.

The targets of Polyergus raids are a related group called the Formica ants. Swooping down on a Formica colony, the Polyergus raiders spray the nest with an alarm chemical that causes the Formica to panic. As they flee, the Formica workers attempt to save their young by grabbing the *pupae* and scattering in all directions. Left to fend for herself, the queen also abandons the nest.

The conquering Polyergus enter the nest and capture the young that have been left behind—sometimes as many as 3,000 pupae—and return to their own nest. Some of the Formica young are eaten, but most are allowed to grow up in their captors' home.

Formica slaves are tricked, rather than forced, into working. As the young Formica workers emerge from the pupal stage, they *imprint* on whatever odors they first encounter—in this case, the smells of a mixed Polyergus-Formica nest. Like captured wolf pups that respond to their human family as they would have responded to their own wolf pack, newly emerged Formica workers respond to their new environment as they would have responded to their own colony.

Like ant workers everywhere, they set about performing the tasks necessary for the maintenance of the colony and the care of the young. All of the essential work of the colony—foraging for nectar and dead insects, feeding and washing the Polyergus queen, workers, and larvae, and defending the nest against attackers—is done by the Formica ants.

The one thing they cannot do, however, is reproduce themselves. As workers, the Formica slaves are sterile. Periodically, the Polyergus workers must raid another colony to replace their aging slaves.

HONEYBEES: A DANCE THAT "SPEAKS"

Honeybee society also consists of a queen, her daughter workers, and *drones*, or breeding males. Although these bees do not weave nests, maintain gardens, herd other insects, drop stones on competing insects, or take slaves, they have generated intense interest for another reason. Their ability to communicate with one another is so amazing that scientists refer to their system of signaling as the honeybee "language."

Like ants, foraging honeybees return to the hive to recruit others to help bring back the food they have found. It is easy to see how ants tell other members of their colony how to locate the food. The scouts simply leave an odor trail behind them on the way back. In order to find the food, all the other ants have to do is follow this trail.

Honeybees travel through the air, however, not on the ground. Not only do smells dissipate easily in the air, but food sources downwind of the hive could not be detected by smell. Moreover, some of the sources of food that bees find are up to a mile and a half (2.4 km) away, much further than the food sources ants recruit to. It is not possible, therefore, for bees to leave an odor trail like ants. Yet, shortly after the foraging bee has returned to the hive, her colony mates pour out and make a "beeline" for the food source.

Since the forager does not lead the others to the food, she must have some way of telling them how to find it on their own. The inside of the hive is totally dark, so it is obvious that she cannot simply point in the correct direction. Moreover, the others know not only in what direction to fly but how far to go. How in the world does the little honeybee communicate this kind of information?

Some of the honeybee's signals are chemical, much like those used by some other insects. When she leaves the flowers she has found, the honeybee marks them with a gland on her abdomen. Once others from the colony reach the vicinity of the flowers, this odor will tell them exactly where to land.

The smells on her body also identify the forager as belonging to the hive. Since she has spent her entire life in this hive, she is covered with its distinctive odor. This smell lets guards at the hive entrance know that she is not an intruder from another hive interested in stealing honey.

However, none of this explains how the forager can direct others to a site a mile or so away. To do that, the little bee must dance.

After passing the guards at the hive entrance, the forager enters the darkness of the hive. Nearby bees, attracted by the scent of flowers clinging to her body, approach the forager to investigate. By offering them samples of the nectar she has collected, she shows them what kind of food she has found. Then, pushing her way through the bees clustered around her, she begins to climb the combs lining the hive. As she climbs, she buzzes and wags her tail vigorously from side to side.

Intensely interested, other bees follow her in the dark, maintaining constant contact with her by touching her with their antennae. At a certain point, the forager makes an abrupt turn and goes back to the spot where she started. Then she repeats her tail-wagging, buzzing climb back up the combs.

Amazingly, it is through this strange dance that the forager tells others where to find the food. Each part of the dance conveys a different message. How hard she wags her tail tells others how good the food is and helps them decide whether making the trip will be worth it. The length of the waggle section of the dance tells them how far away the food source is. This is important in letting the bees know not only how long they must fly but how much fuel, in the form of honey, they will need for the trip.

The most complicated part of this dance code, however, is the direction the forager climbs when she waggles. This is the part that tells the other bees the direction they must fly to find the food. When the food source is in the direction of the sun, the waggle part of the dance is performed straight upwards. By contrast, when the food is located in the opposite direction, that is, directly away from the sun, the forager wags her tail as she travels down the combs. Food located to the right of the sun's position is conveyed by a waggle dance that moves horizontally to the right, while a waggle dance that moves to the left tells the bees that the food is located to the left of the sun.

By combining the information learned from various parts of the dance, the bees can tell in which direction they must fly, how far they must go, how much fuel they need to take, and whether the food is good enough to justify making the trip. Because there are many different messages encoded in the dance, the bees must follow the forager several times before they know exactly what to do. But once they take off, most will reach the food successfully with little trouble.

In other species as well as bees, knowing the rules of a particular animal society makes it easier to figure out what individuals are doing and why. From frogs croaking furiously in a pond to penguins performing bizarre nest rituals, understanding the social organization of any species makes it possible to untangle behaviors that might otherwise seem very confusing. Best of all, an animal's social rules give us an idea of what it might actually *feel* like to be an individual in that society. If you were allowed to spend a month in one of the animal societies discussed in this book, which would you choose?

Glossary

crèche a day nursery for infants

drones male honeybees; unlike females, they do not sting and they gather no honey

habitat the place or the type of site where a plant or animal normally lives or grows

hierarchy a group of animals organized in a chain of authority, with each individual in the chain having more power than the individuals beneath it

imprint to make an indelible first impression that affects an animal's subsequent behavior or connection with other animals

migration the periodic movement of animals from one region or climate to another for feeding or breeding

monogamous animals that have only one mate at a time

nestlings birds that are too young to leave the nest

nomadic animals that roam from place to place in search of food

pack a group of animals that hunt or run together, especially a group of wolves or dogs

polygamous animals that have more than one mate at a time

predation the killing of one animal by another for food

savanna a wide area of meadowlike land, a treeless plain

selective breeding the practice of allowing only individuals that have desired traits to breed

species a group of animals that breed among themselves but not normally with members of other groups

tundra a vast, level, treeless region with an arctic climate and vegetation found in northern Europe, Asia, and North America

For Further Reading

Gormley, Gerard. *A Dolphin Summer.* New York: Taplinger, 1985.

Gravelle, Karen, and Ann Squire. *Animal Talk.* New York: Julian Messner, 1988.

Kaehler, Wolfgang. *Penguins.* San Francisco: Chronicle Books, 1989.

Lopez, Barry Holstun. *Of Wolves and Men.* New York: Scribner, 1978.

Moss, Cynthia. *Portraits in the Wild: Animal Behavior in East Africa,* 2d ed. Chicago: University of Chicago Press, 1982.

Sattler, Helen Roney. *Sharks: The Super Fish.* New York: Lothrop, Lee & Shephard, 1986.

Squire, Ann. *Understanding Man's Best Friend: Why Dogs Look and Act the Way They Do.* New York: Macmillan, 1991.

Index

About the Author

KAREN GRAVELLE has a Ph.D. in Animal Behavior and works as a writer and photographer in New York City. To earn her degree, she spent many months observing animals in the southwestern United States and the West Indies. Dr. Gravelle has followed army ant colonies as they raided their neighbors and watched countless lizard battles over turf. She is the author of *Fun Facts about Creatures, Animal Talk*, and the Franklin Watts book *Lizards*.